YELLOWSTONE LAKE

JOIN THE EXPEDITION!

By
Sory Marocchi
with Scientists
Tony Remsen & Val Klump

Published in the United States by
Hammockswing Publishing
5017 North Shoreland Avenue
Whitefish Bay, WI 53217-5540

Marocchi, Sory
with Tony Remsen and Val Klump
Yellowstone Lake: Join the Expedition!

ISBN 0-9706606-0-X

Design by
Sory Marocchi & Milestone Marketing
Cover design by Jason Robson

FOREWORD

Yellowstone Lake: JOIN THE EXPEDITION! reminds me of many of my own at sea adventures using undersea robots to explore the underwater wonders of the deep sea. It brings the excitement of scientific exploration to young people and the general public in a most compelling way.

You can pick the book up and open it to any section to find an intrepid scientific explorer describing his or her efforts at unraveling the secrets of one of the most unusual landscapes on our planet.

These expeditions to Yellowstone Lake bring together scientists with many diverse interests all committed to better understanding the geysers and fumaroles, mudpots and hot springs that lie hundreds of feet beneath the surface, beyond the reach of traditional SCUBA divers. Using small, remotely operated vehicles or ROVs, they probe fascinating geologic features that rest hidden in the watery domain of one of the world's largest active volcanic hot spots.

And most importantly, what they learn may help us better understand the origins of life on our planet, as well as help us plot strategies for finding life elsewhere within our Solar System or beyond.

If you take a few moments to explore this amazing lake, you will want to learn more about our planet and its origins.

Dr. Robert Ballard

TABLE OF

CONTENTS

YELLOWSTONE LAKE

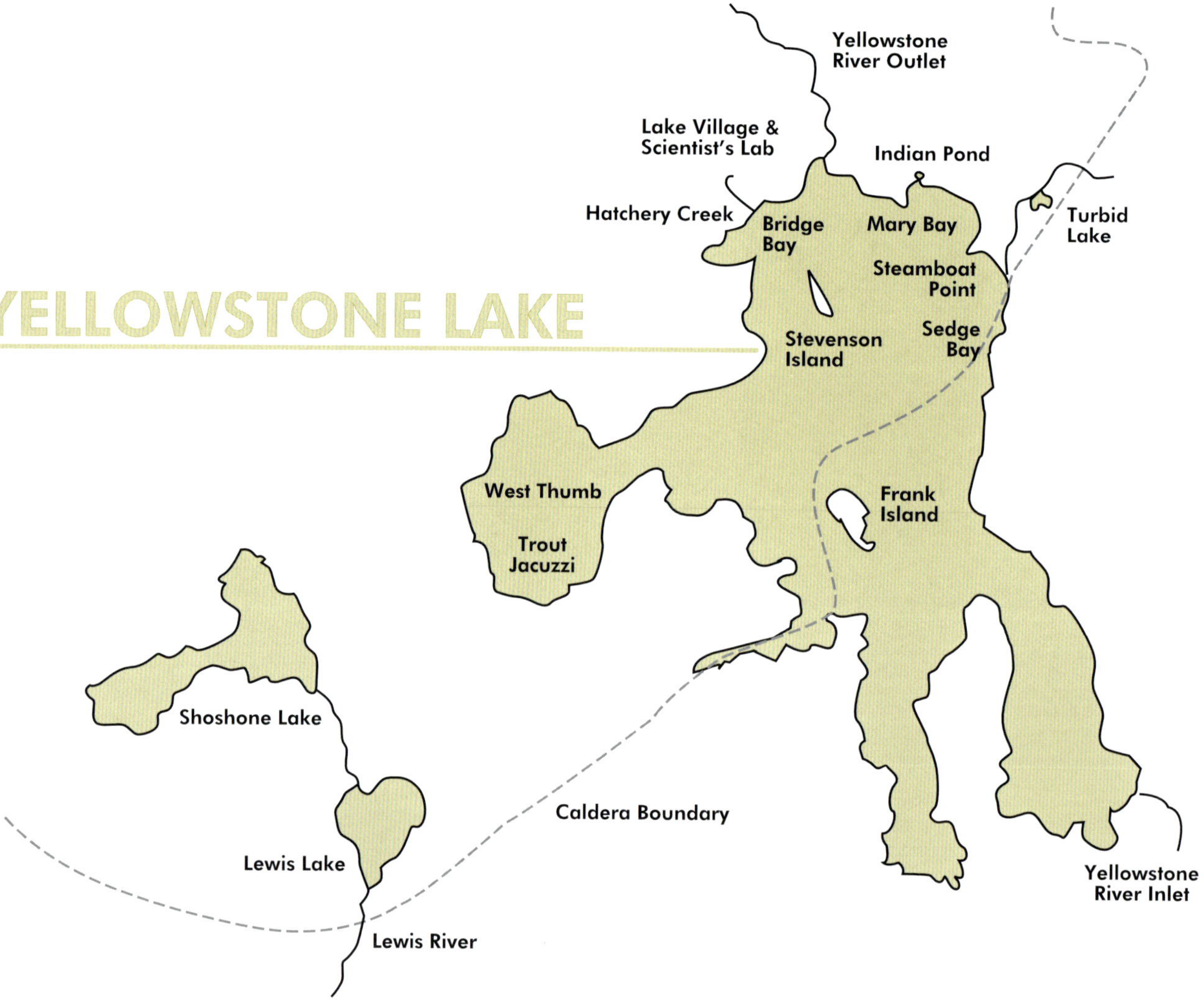

GETTING THERE IS HALF THE FUN

This expedition, and the discovery of one of the most fascinating places on earth, began in a wastepaper basket at the Great Lakes WATER Institute of the University of Wisconsin-Milwaukee. It was late afternoon when one of our graduate students came into my office and pulled an RFP (Request For Proposals) out of the 'circular file' where I had just tossed it. "I don't have time to spend on such a small project so far from Milwaukee," I explained to Mike, but he -- like all good graduate students -- didn't take the word of his advisor. As it turned out, his simple "Hey, this sounds interesting" was the understatement of the year.

Mike persuaded me to respond to what was, undeniably, an intriguing problem: a lake that appeared to be aging backwards, like Merlin, the legendary magician who grew younger as years passed. Normally, the water in lakes becomes increasingly concentrated over time, as salts, soils and debris wash off the land during each spring thaw and rainstorm. Admittedly sketchy data suggested, however, that for some unknown reason, Yellowstone Lake was doing just the opposite. Someone had even extrapolated the trend as reaching distilled water within a decade! Now, a distilled water lake is impossible (some things are not possible, and this is one of them), but even the suggestion was unusual enough to warrant a second look.

Consequently, several months later, two of my colleagues and I found ourselves headed west on Interstate Route 90, driving a high mileage university-owned station wagon, and pulling a U-haul trailer full of scientific paraphernalia and sampling gear. Our mission was to collect sediment cores (vertical sections of mud) from the bottom of the largest high altitude lake in North America. In lakes, sediments (the stuff that makes mud puddles muddy) slowly sift their way to the lake floor and accumulate bit by bit, year after year, until impressive amounts of mud build up to maybe tens of meters deep. The beauty of this process is that a sediment core is like a tape recorder of the history of the lake and its watershed. The deeper you penetrate into the mud, the further back in time you go. Were there clues about Yellowstone Lake's aging process recorded in its sediments?

It was to find this out that we first journeyed to Yellowstone in 1983. One of the first things we discovered was that our old flat-lander station wagon, not keen on towing a trailer some 1400 miles, was even less prepared to pull it up and down entire mountain ranges in the process. We suffered overheating brakes, overheating engines and overheating scientists, but we made it there and back, barely. What we saw was enough to lure us back the next year, and the year after that, and the year after that...

Seventeen years later, we are still packing up trailers and trucks full of gear and heading west to explore one of the world's aquatic gems and the mysterious living world beneath her waves.

Launching a complex scientific expedition in a fairly remote area is not a simple task. First of all, there is the gear. You might think your mom takes a lot of stuff when she travels, but she has nothing compared to a bunch of scientists. Our single station wagon and U-haul combination, first expanded to include a full size van, 14′ trailer, and pick-up, is now replaced by a truck designed to hold a 6 room house full of furniture. Once emptied, this doubles as our field laboratory.

Over the years we have had our share of "transport problems". Besides overheating things in general, we've had trailer tongues break, trailer wheels fall off, trailers fall off, electrical problems (always), fuel pump problems, blown mufflers, a blown engine -- in fact, I don't recall a year in which something has not gone wrong. This being a low budget operation, motel bills are to be avoided, so 30 hours' straight driving is our planned itinerary. Summer traveling is cooler at night anyway, and covering the distance is quite possible with 3 or more drivers. I should emphasize the more, because to launch this kind of expedition with boats, divers, underwater robotics, gadgets galore, instruments and experiments, takes a team.

Over the years more than 40 people, scientists and students from all over the country, have contributed to the success of our explorations.

This book is our attempt to have you, the reader, join our expedition and share with us the unseen wonders at the bottom of Yellowstone Lake. Along with our scientific discoveries, we have each described a personally memorable moment from our Yellowstone experiences. As an honorary member of our team, we've provided you with an official expeditionary postcard, which we hope you will use to share some of your own observations in Yellowstone. Let us know what you find interesting, and we'll check it out. Thanks!

Val Klump
UW-Milwaukee
Great Lakes WATER Institute

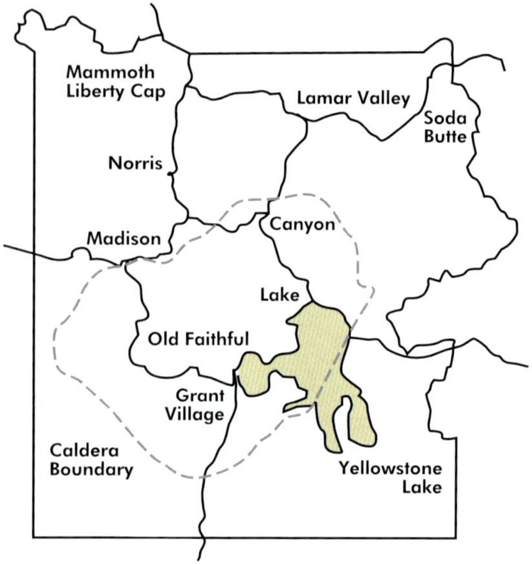

IN THE HOTSEAT

Most visitors to Yellowstone probably don't realize that they are walking around on top of an ancient volcano in one of the most tectonically active "hot spots" on earth. Some 600,000 years ago, in one of our planet's largest known volcanic eruptions --1,000 times that of Mt St Helens -- a huge chunk of earth exploded into the atmosphere, scattering as far away as Kansas, and leaving behind a crater some 30 miles across, which now forms the heart of Yellowstone National Park.

Within this crater are two "resurgent domes" or swellings in the earth: one around Old Faithful and the other at the northern end of Yellowstone Lake at its Yellowstone River outlet. Episodically, the ground in these areas is known to rise up to several inches in as short a time as 3-4 weeks, tilting the lake's outlet upward and increasing natural fluctuations in the lake level, which result annually from snow melt and evaporation. (Old beach ridges, such as those which can be seen near the Fishing Bridge Visitor's Center, provide evidence of even greater variations in the lake's level which have occurred over hundreds, and even thousands, of years.)

CONCERNING THE GEOLOGY OF YELLOWSTONE LAKE

Yellowstone Lake straddles the edge of the old volcano (the caldera rim), and its waters may cover one of the most tectonically active areas in the Park. Though so mild as to be unnoticeable to most visitors, micro earthquakes continue to shake this region many times each day.

In 1985 University of Wisconsin-Milwaukee scientists made the acquaintance of geologists Paul Morgan, David Blackwell and Matt Kaplinski as they were measuring heat fluxes under the lake in order to study continental drift over the Yellowstone hot spot. It was while collecting data near Stevenson Island that the three men noted inexplicably hot readings with their temperature probe.

A grant from the National Geographic Society allowed Matt to return in 1986 and begin to chart the bathymetry (map the bottom) of the lake. His studies indicated unusually steep depth contours near Stevenson Island, leading our scientists from the University of Wisconsin-Milwaukee to return to explore the area in 1987.

Park records have subsequently been changed from portraying a deepest spot of 310 feet in the West Thumb area of Lake Yellowstone to indicating a (current) deepest official measurement as 390 feet just off Stevenson Island. Thus was pinpointed for the first time an area on the lake bottom which continues to fascinate and challenge scientists to this day.

In the summer of 1988 when the fires were raging, Joe Hazel and I were profiling down south of Frank Island. We thought we were the safest people in the park – surrounded on all sides by water in the middle of the inferno. Around midday, I smelled something funny and poked my head out from the cabin and asked Joe if he smelled smoke. Joe informed me that the whole !!!!!!!!! park was on fire, of course he smelled smoke. Then we looked back by the engine cowling and smoke was billowing out of a pile of stuff we stashed there for the day. Among the items were two six gallon cans of gas! Reacting quickly, I did the first thing that came to mind and dashed back into the cabin yelling "!!!!, we're still on line. I've got to save the file!" Meanwhile Joe ripped at the pile of cloths, coolers, generators, and gas cans, calmly grabbed the fire extinguisher and put the fire out. It turned out that our DC bilge pump shorted out. The wires from the pump to the battery melted down and started the straps of the battery cover on fire. And that's how we almost blew ourselves to smithereens.

Matt Kaplinski
Northern Arizona University

As guest scientists in the park, we were allowed the privilege of diving in Yellowstone Lake. Even after park approval, though, we notified officials daily as to where and when we would be diving. We kept in close contact with park personnel and other members of our research team via 2-way radios. Following PADI (Professional Association of Diving Instructors) dive charts and guidelines for high altitude diving, we limited ourselves to depths of about 30 feet. Fortunately there was an abundance of sites within this shallow depth, and water clarity was excellent.

THROUGH THE LOOKING GLASS

Yellowstone National Park is one of the coldest areas in the lower 48 states, and the surface of its large lake typically lies under up to 3 feet of ice from December to June. In the summer of 1985 University of Wisconsin-Milwaukee scientists returned to the Park for the third time, eager to begin exploring near shore areas of Sedge Bay which, interestingly, were reported to remain ice free throughout the winter.

Taking advantage of a large flat rock which protruded from the lake surface about 12 feet from the shore of Sedge Bay, team members began (with permission from the Park Service) to use it as a SCUBA diving platform, thus encountering their first glimpses of a spectacular underwater world.

Several kinds of hydrothermal activity were immediately evident. Mesmerizing streams of bubbles could be seen escaping from the sandy bottom, while in other places shimmering curtains of hot water rose from both open sand and rocky clefts. Beneath 6 to 20 feet of water, divers observed dense patches of seaweed-like plants rooted to the floor of the lake. Expanses of aquatic moss looked almost like lawns, while some plants sported short sharp leaves reminiscent of evergreens. Flora flourishing near the hot vents appeared, incongruously, to be covered with snow, an effect which was later attributed to the presence of sulfide-oxidizing bacteria which colonized these plants while extracting energy from the hydrogen sulfide escaping from the lake bottom. Divers could detect the familiar "rotten egg" smell of this gas even underwater, through their masks.

CONCERNING SCUBA DIVING IN YELLOWSTONE LAKE

The most amazing sight which greeted the explorers was what appeared to be a two to three foot wide, multicolored carpet undulating along a fissure in the rocky lake bottom for as far as 100 feet. This mat, so densely packed as to be trapping pockets of gases beneath it, later turned out to be composed of photosynthetic and chemosynthetic bacteria coexisting together in a spectrum of colors which ranged from black and white to green and orange, with a purple swath down the center. While smaller mats have been noticed on a routine basis in subsequent explorations, such a huge spectacular one has been seen only once.

Though fascinated by what they had observed, the scientists were also frustrated by their inability to explore deeper than 30 to 45 feet for long periods of time, due to the near freezing temperature of the water as well as decompression problems, which become increasingly dangerous at high altitudes.

Bacterial Mat

I remember one field of fumaroles in Sedge Bay as being particularly dense and vigorous. The sound of the bubbles bursting as they streamed upward was even audible over the noise of the SCUBA gear. I had hoped I could feel their gentle caress, but the heavy wet suit disallowed any sensation. As divers in an otherwise cold lake, however, we did relish the flows of warmth from the underwater hot springs. We typically dove twice a day, keeping in mind that the weather changed quickly and often.

Lorie Buchholz
of the WATER Institute
now lives in Denmark
with her family

Measuring temperature and dissolved oxygen in Sedge Bay where white sulfur deposits on plants indicate the presence of hot vents.

Fumaroles, hotsprings, bacteria mats and lush plant life create magical underwater scenes in the more shallow areas of Mary and Sedge Bays.

In July of 1989, as the sub was down exploring the lake bottom near Stevenson Island, we realized that our fathometer was reading 375', much lower than the officially charted maximum depth of 310' for Yellowstone Lake. With great excitement we watched towering, sheer cliffs and hot water vents pass the camera lens as the ROV sank deeper and deeper, until it was more than 400' below the surface.

But then the video picture stopped moving, and though not everyone realized it right away, my little sub had in fact become stuck, wedged in an increasingly narrow trench. With my joystick, I tried gently to rock the vehicle back and forth in all directions, but every effort seemed only to worsen the already bad visibility, as the sub jostled, jerked, and lurched into walls and sediment banks. It probably only took minutes, but it seemed more like hours to me before I managed to work the sub free.

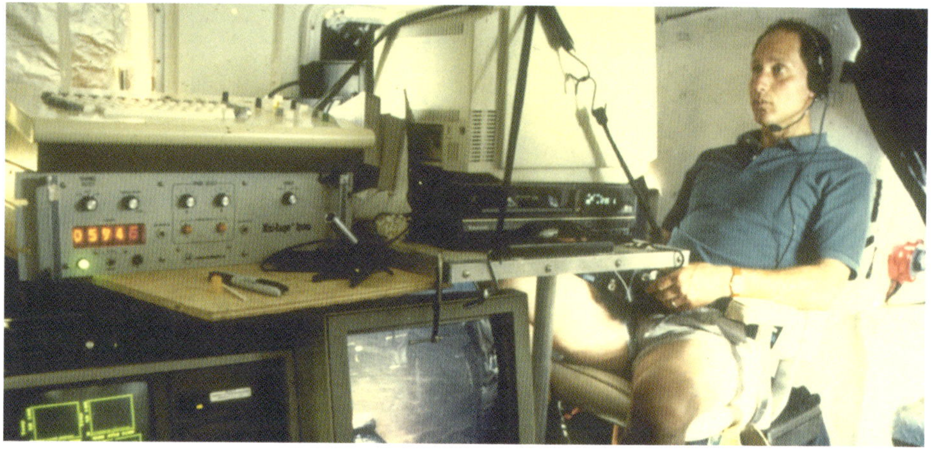

LIGHTS, CAMERA, ACTION!

For their first three summers of exploring the hydrothermal vents under Yellowstone Lake (1984-1986), our scientists were confined visually to what could be seen while SCUBA diving in relatively shallow areas of Sedge and Mary Bays. Even wearing cumbersome dry suits, diving much deeper than about 30 feet in such cold waters and at such high altitude was not practical. Their curiosity, however, was piqued by mounting evidence that hot vents existed in the deepest parts of the lake off Stevenson Island, and our researchers set out to find the technology which would allow them to inspect all areas of the lake bottom.

Having experienced diving in manned submersibles in Lake Superior, the team realized that the use of such a large scale (requiring a substantial mother ship) and expensive ($10,000-$40,000 per day) system was impossible on Yellowstone Lake, where the hot vents and precipitous topography of the lake bottom would, in any case, make diving extremely dangerous. Accordingly, a modest grant of $4,900 was sought from the National Oceanic & Atmospheric Administration for a week's use of a Remotely Operated Vehicle or ROV, which would carry a camera (but not human beings) about underwater. Thus in July of 1987 Dave Lovalvo of Eastern Oceanics brought his little 3 foot long "Yellow Submarine", much technical expertise, a tireless work ethic and cheerful, can-do attitude to the Yellowstone research effort for the first time.

CONCERNING THE USE OF ROV's IN YELLOWSTONE LAKE

Dave's first vehicle was small, cute and simple. One person could easily lift it out of the water. Technically it was a member of a new class of "suitcase ROVs" which were little more than motorized cameras "flown" underwater on the end of a long cable, or umbilical. Easy to pack, transport and deploy off a 28 foot work boat loaned by the US Fish & Wildlife Service, this vehicle ("Mini" was her name) proved ideal for exploring the bottom of Yellowstone Lake. And explore Mini did! Unusual plants, animals and geological formations filled the video screen. There was only one problem: the scientists could see a whole new world of curiosities, but couldn't take any samples!

So the team went literally "back to the drawing board" in order to outfit Mini with arms and hands as well as eyes. Today the ROV is no longer so "mini": she weighs nearly 200 pounds. Yet at only 4 times her birth weight, she carries 40 times the equipment. Resembling a large tinker toy, the current sophisticated underwater sampling platform is constructed of light-weight aluminum rails and bristles with tools:

- a 35mm still camera and strobe light
- sonar and a magnetic compass for navigation
- a retractable basket & scoop for collecting samples of anything that looks interesting
- a 3 function manipulator arm for collecting & storing larger objects
- a computer-operated "sipper" system which collects & stores water samples in syringes
- a Hydrolab™ recorder which continuously measures depth, temperature, conductivity, pH & chlorophyll
- a 3 chip broadcast-quality video camera like that used to film Titanic .

As we go to press, ROV work is well into its second successful decade of use under Yellowstone Lake. Each year refinements continue to be made to Dave's equipment as scientific curiosity drives technology to the cutting, challenging and often fragile edge of what becomes possible. Many late nights are spent repairing and refining the sensitive and valuable equipment for another challenging day of sampling and exploration on the lake bottom. Even with floodlights mounted on the front of the vehicle, remotely maneuvering it underwater has been compared to "exploring a dark room with a flashlight" and "driving through mountainous terrain in a snowstorm".

On this compact collection of – temporarily stuck – plastic and metal rested my whole financial future. Though insurance would have covered most of the cost of its component parts, the sub was one-of-a-kind and could only have been painstakingly rebuilt by hand over a period of weeks and months, time I didn't have. The sub and I were already under contract to perform research tasks in other places, for other scientists, soon thereafter.

Businessmen will understand that if I ever don't show up in the contracted place, at the right time, with the requisite working equipment, future jobs will go to someone else. I could even be sued for breach of contract. Especially in the summer months, when weather permits, scientists often want to work around the clock to collect maximum amounts of information from expensive equipment. Time and such opportunities can be neither insured nor replaced.

Dave Lovalvo
Eastern Oceanics

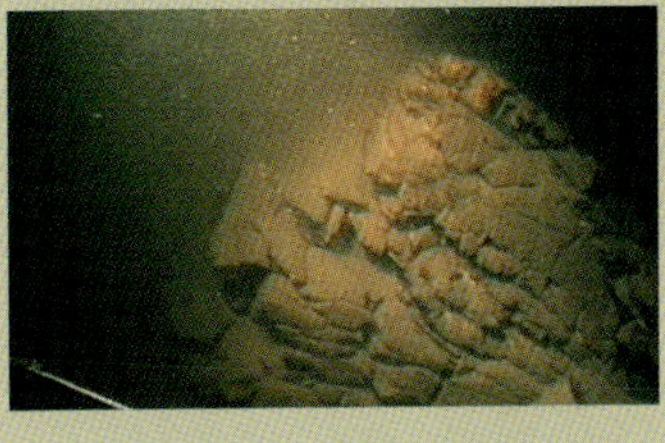

"Mini" could originally be lifted in and out of the water by one person, but as the years passed, many refinements were added. The vehicle is reconfigured every year according to the work planned and new technical possibilities.

Less noticeable than the orange floats, a vent cap (above) and syringe (below) collect bacteria and water samples.

"Mini" in 1990

1. Pressure & Temperature Sensors
2. Flood Light
3. Strobe Light
4. Electric Thruster Monitor
5. 35 MM Still Camera
6. Floodlight for Video Camera
7. Video Camera in Pressure Housing
8. Sampling arm & motor
9. Temperature Probe
10. Compass helps pilot navigate

The ROV equipped with a 3 chip camera in 1997.

An early pumping system was used to continuously collect vent water.

Late-night hours in the shop revamp the complex ROV, not just yearly, but often on a daily basis. The vehicle is now so large that a winch and strong handlers are needed to hoist her into and out of the water.

One would think that after working with the federally funded National Undersea Research Program for twelve years, one could begin to understand and predict the way the government funds science, but not so! Every year the budget process is a roller coaster ride filled with surprises that can leave one dizzy and nauseous. It's no wonder, given the complex path that funding support for science must follow.

At NURP, a proposal for exciting research (such as that in Yellowstone Lake) winds its way from a mail review, to a rigorous, regional panel review, to internal review at the national level. The year before, NURP had to submit its budget to the National Oceanic &Atmospheric Administration, which forwarded priority requests for review at the Department of Commerce and the Office of Management and Budget, before being rolled into the President's budget for that year. That's the easy part.

That budget is shipped off to Congress, which has meanwhile set its own targets in Budget, Authorizing and Appropriations Committees. Mind you, the House of Representatives and the Senate both develop budgets, which must be resolved and signed by the full House and Senate, before being sent back to the President. Are you with me so far?

PAYING THE PIPER

Support for this research has been a partnership of universities, federal agencies and private foundations. Funding over the years has come from the National Park Service (1983), the National Geographic Society (1985-86), the National Undersea Research Program (1987-90, 92, 94-97, 2000), the National Science Foundation (1998-99) and the Dreyfuss Foundation (2000). Grants have ranged from a few thousand dollars for one or two years to as much as $100,000 per year. The US Fish and Wildlife Service and the National Park Service have additionally provided invaluable "in kind" support: laboratory facilities, boats, dock space and lodging. Because laboratory and living space in Yellowstone is very limited, the Lake Village's fisheries group's willingness to accommodate us during their own busy summer schedule is a testament to their patience and fundamental drive to understand this amazing lake. We have also enjoyed the support of our home institutions, the Graduate School of the University of Wisconsin-Milwaukee (for most of us) and our sister institution of higher learning in Milwaukee, Marquette University.

A research project has several stages, each of which is critical to its success. First is the set of ideas, questions and hypotheses ("How does this work?" or "What lies on the bottom?") which are generally triggered by observation (noticing bubbling hot springs near the shore, steep thermal gradients in the mud at depth or inexplicable fathometer recordings). Typically, as university based scientists, we submit our ideas in the form of a research proposal (approximately fifteen pages in length) to a funding agency, as do other scientists from all over the country, seeking to do work all over the world. Proposals are then sent out for review by other experienced scientists, who assess their merit in terms of the objective (knowledge to be gained), the probability of success and the practicality of the approach. Agencies always have more worthy proposals than money available, and so must, with difficulty, prioritize the research projects. A panel of scientists and agency administrators usually meet to rank the submissions. Only 10 to 30% of proposals generally receive grants.

CONCERNING THE FUNDING PROCESS

Being lucky enough to receive a grant means that the details of the research must then be planned. If travel is required (as it is for the work in Yellowstone), logistics such as transportation, food, lodging, laboratory space and ship time must be worked out. New sampling tools are designed, built, tested and reconfigured. This is not a casual effort.

Eventually, everything needed from Milwaukee (including people, instruments, dive gear, spare parts, tools, etc) is transported to the research site. Working in remote areas may mean that a single piece of forgotten or broken equipment results in the loss of hours or even days of work time. Over the years, as we've brought more equipment and the ROV has grown larger, we have necessarily improved the working platform of the R/V Cutthroat, rewiring the electrical system and adding more anchors, better winches, bigger generators, an ROV launching system, even a port-a-potty! Video tapes and dive logs have been meticulously kept and archived by NURP. Improvements are constant. It is hard to remember the early days with little "mini" and a simple temperature probe.

Finally, back in our offices in Milwaukee, results are tabulated and findings published in scientific journals after being reviewed by colleagues for clarity, value and content. This book (intended for the general public) is different from anything we have published before, and is made possible through the support of the Dreyfuss Foundation, NURP and the UW-Milwaukee.

Of course, with research like ours, we must confess that Mother Nature has the final say. One summer, the weather was so bad for the whole time we were in the Park, that dives were confined to a few early morning hours on a lee shore. As you can imagine, it can be very frustrating to gear up people and equipment, only to sit at the dock while the winds howl.

We are used to the Great Lakes, whose storms are famous, but have learned that the weather on Yellowstone Lake can also change within minutes. Caution and common sense are essential. We maintain a regular schedule of radio contact with the Park Rangers, whose unceasing efforts to keep all mariners safe is a truly wonderful service. We have ourselves helped rescue capsized canoers, warned others of bad weather and given the occasional tow, which is only fair after receiving one ourselves.

We are thankful for the privilege of studying one of nature's most beautiful and intriguing lakes and could not have done so without the people and agencies who have "joined this expedition" and made it possible.

Now, let's add time schedules to the circuitous route the funding has to follow. The federal fiscal year (Oct 1st to Sept 30th) determines when NURP receives its funding. It may be April before regional centers receive support slated Jan 1st to Dec 30th. Funding is then forwarded to university systems on a state fiscal year of July 1st to June 30th.

As you can see, it's a wonder that such exciting, imaginative research as described in this book ever gets supported at all! Just as there are a thousand details for the scientist to stay on top of, so must those who support science be wary not to let anything "fall between the cracks" as the funding cycle turns each year. We are thankful for the patience and perseverance of the scientists who deal with devilish details before and after their project gets supported.

Ivar Babb, Director
NE Atlantic & Great Lakes
National Undersea
Research Center

The horizon shimmers from the heat like a country highway in midsummer. Amid the barren fields, occasional stumps and crevices smolder. From some, flakes of whitish ash-like material drift lazily upwards, buoyed by the heat.

Quickly mopping the sweat from his brow, the driver peers intently ahead. The target, an awesome spire, is in sight, but some rugged terrain lies between it and his vehicle. His hands flash across the control panel, trying to keep his sluggish vehicle from crashing into a crevice wall. A cloud of dust swirls upward, obscuring the view for a moment. As the particles settle, the driver gasps. Annealed like ceramic, but from no human kiln, the towering monument to subterranean fury, fully 25 feet high, looms directly ahead!

SUBTERRANEAN COOKERY

The unusual proximity of magma -- hot molten rock -- to the surface of the earth in Yellowstone gives the region its unique geography. Water filtering downward through porous soil or fissures in rock becomes heated far above its normal boiling point, then travels upwards to emerge as:

Hot springs	where hot water flows continuously to the surface of the earth.
Fumaroles	where hot gases escape from underground.
Geysers	such as Old Faithful - from which jets of hot water and steam emerge, episodically or at predictable intervals.
Mud pots	where smaller amounts of heated water containing chemicals such as sulfuric acid react with the soil, sometimes creating the various colored "paint pots", depending upon which minerals are present in both soil and water.

From their fist visit, scientists began to hypothesize that such hydrothermal venting was also taking place beneath the water of Yellowstone Lake. Their suspicions were soon confirmed.

CONCERNING GEOTHERMAL ACTIVITY UNDER THE LAKE

In 1984, researchers began sampling from these underwater vents. Their first approach was simple and direct: SCUBA divers stuck the end of a 1/4 inch diameter piece of flexible tubing in a vent opening and pumped hot water directly into analytical instruments floating 12-20 feet above them aboard a small boat. These early (and somewhat crude) experiments yielded dramatic results: chemicals/minerals in the vent water were at concentrations 20 to 30 times greater than that of surrounding lake water, and up to hundreds of times greater than what is generally found in alpine settings. It was immediately clear that the input of these geothermally produced nutrients could be having a significant effect on the ecology of Yellowstone Lake.

Studies are now being undertaken which will enable scientists to accurately correlate water temperatures with mineral/chemical concentrations. "Chemical geothermometers" indicate that under great pressure, water may reach 200 degrees Celsius or 414 degrees Fahrenheit without being turned to steam. It has been found that higher temperatures often mean that greater amounts of dissolved nutrients will be present in the waters escaping from hydrothermal vents.

Our team could not help realizing that here in Yellowstone they had found smaller, much more accessible, freshwater versions of the famous deep sea hydrothermal vents, which erupt from chimney-like forms beneath the Atlantic and Pacific Oceans. The water flowing from the floor of Yellowstone Lake also contains concentrations of minerals, which apparently react with lake water and sediment to form the pipes, spires and chimneys observed through the ROV's cameras.

Working under Yellowstone Lake with the ROV can seem like an adult version of Nintendo, but this exploration is not a game; the victories and defeats are all too real.

So why are we doing this? The development of the sophisticated technology required for such underwater exploration allows us to more easily reach new frontiers--from the bottom of Yellowstone Lake and the deep sea, to the robotic exploration of outer space. These bubbling geothermal pools and vents may provide clues as to how primordial life began on earth. The chemical and biological activity they trigger may give us insight into the potential for life on Mars or the moons of Saturn.

Maybe most importantly, understanding how Yellowstone Lake "works" will help us protect and preserve this natural wonder for future generations.

Russel Cuhel
aka "Dr. Cool"
UW-Milwaukee Great Lakes WATER Institute

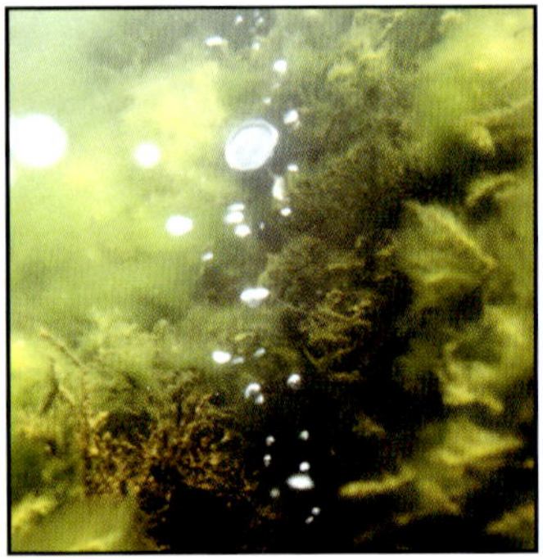

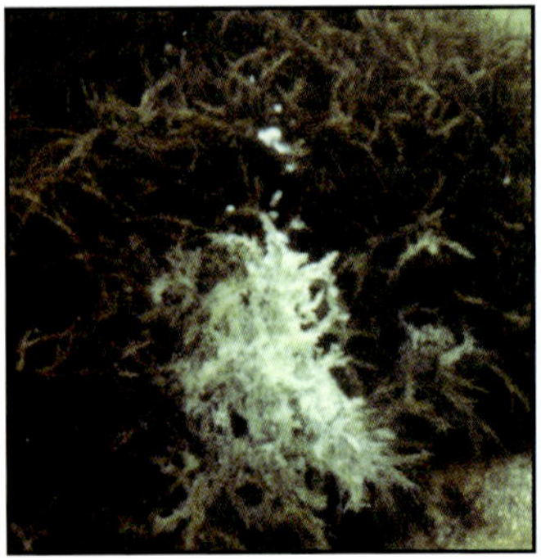

CURTAIN CALL

On a calm day, from the shore of Lake Yellowstone's 'West Thumb', park visitors can see evidence of an underwater geyser. It was Park naturalist John Dahlheim who first alerted us to vigorous disturbance of the water's surface which occurs there about every 25 minutes.

During the summer of 1990, taking advantage of the ROV technology available to us, we piloted the 27 foot Research Vessel "Cutthroat" to what we hoped might be the right spot, and lowering the submersible with its camera, discovered a narrowing cavern on the lake's bottom about 20 feet below the surface. We waited for action...

Mary Bay Sponge

Yellowstone Lake is more productive, or full of life, than most high altitude alpine lakes. Not only are photosynthetic processes supporting life which derives its energy from sunlight in the upper 30 meters (98 feet) of the water column, but amazingly, organisms are also growing, reproducing and thriving in total darkness at the lake's bottom, up to 400 feet below the surface.

How is this possible? In a process called chemosynthesis, microscopic bacteria use chemicals dissolved in the water flowing from hydrothermal vents as an energy source. Straight from the vents, the hottest water is richest in dissolved chemical nutrients, but is, in fact, too hot for the survival of most life forms. It is when this water (which would be boiling at surface air pressure) begins to mix with the cold (~4 degrees Celsius, close to freezing year round) bottom water, and warms to about 30 degrees C, that plants and animals begin to flourish in the combination of warmth and nutrients. There is evidence that in Yellowstone Lake, chemosynthetic processes occurring near the underwater vents can produce nearly as much energy as the photosynthesis going on closer to the surface.

Light from the ROV illuminates the darkness surrounding a deep vent.

CONCERNING FOOD CHAINS IN YELLOWSTONE LAKE

Thus food chains thrive at all depths in the lake. At its dark bottom, the "lowest" members, or sources of "primary production", are chemosynthetic bacteria, which are eaten by zooplankton (microscopic animals), sponges, snails, and a variety of worms. Higher in the water column, where there is light, photosynthetic algae feed the zooplankton.

The zooplankton, which can be so prolific as to be described as "swarms", are eaten by fish. Small fish support bigger fish, which are consumed in their turn by pelicans, osprey, eagles and even bears, who venture into the shallower water of rivers which flow into the lake. Humans also function as predators in the food chain of Yellowstone Lake, but confine their fishing within limits set by Park authorities in order to safeguard the balance of the Lake's ecosystem.

Safeguarding that balance is not easy. Park authorities have recently become aware that Lake Trout have been introduced -- by accident or design -- into Yellowstone Lake and threaten its native population of Cutthroat Trout. Lake Trout are aggressive predators, as evidenced by the netting of a 17″ Lake Trout with a 12″ Cutthroat actually in its mouth! In the words of one fisheries biologist: "Lake Trout are really tough on Cutthroat. They just beat the heck out of them, chase 'em down, eat 'em!". Indeed, the danger to the Cutthroat has recently been judged to be so acute that a program to remove the aggressive Lake Trout from Yellowstone Lake via gill netting is now being implemented.

Suddenly, the water all around the boat roiled with turbulence! Watching the video screen located in the RV Cutthroat's cabin, we were surprised and somewhat puzzled to observe a large Cutthroat trout swimming rapidly back and forth through the erupting bubbles. Because this behavior was repeated several times, we concluded that – far from being an accident or coincidence – the fish was deliberately waiting to feed on the clouds of small crustaceans, insect larvae, and bacterial mat material stirred up by the periodic turbulence. We now refer to this underwater geyser as "The Trout Jacuzzi".

Carmen Aguilar
UW-Milwaukee Great Lakes WATER Institute

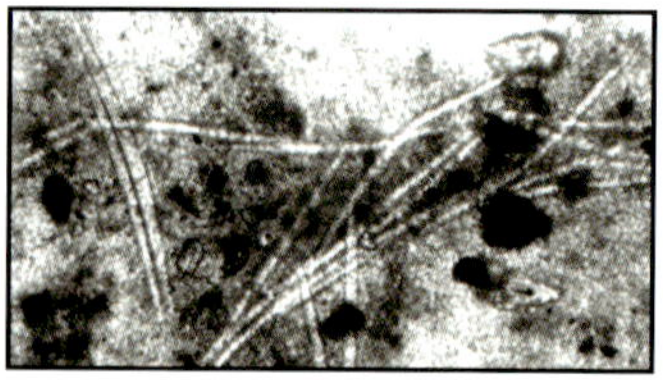

Sponges (above) and diatoms (below) at different magnifications.

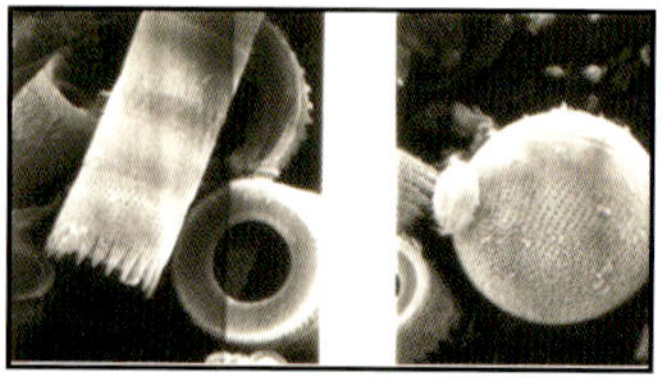

A video camera's blurred image of a Cutthroat trout feeding on small crustaceans near West Thumb's... "trout jacuzzi"

Two pelicans fishing for Cutthroat in the lake's Southeast Arm.

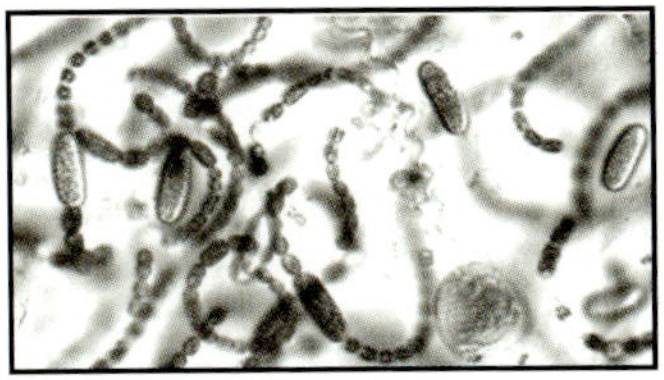

Algae (above) and bacteria (below) as seen through microscopes and with the naked eye.

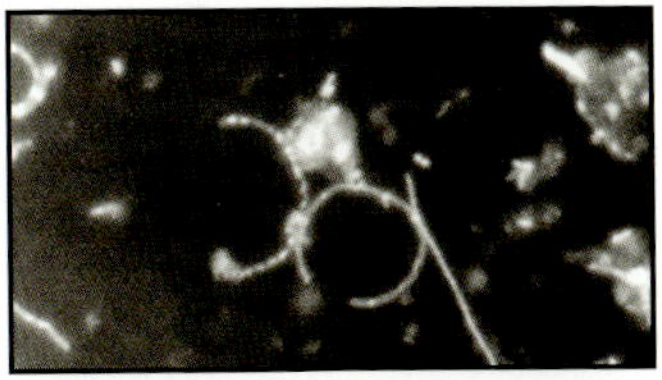

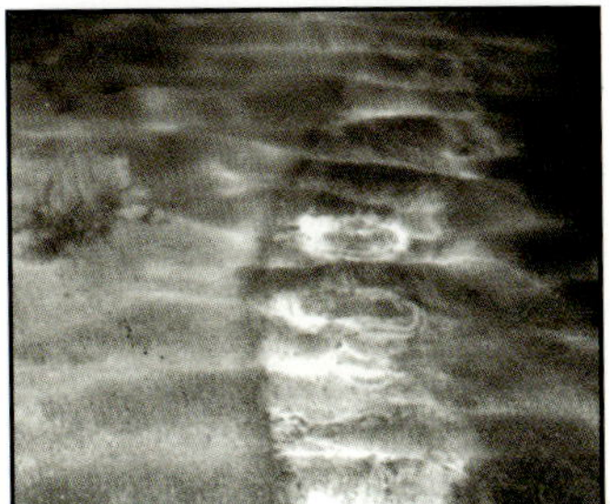

One's first visit to Yellowstone is invariably awe-inspiring, but I think my one winter visit is probably my most memorable experience so far of the great Park.

LINE OF SIGHT

1987 was the first year Eastern Oceanic's miniature unmanned submarine was used to explore the depths of Yellowstone Lake. The benefits of using the sub's robotic cameras were immediate and breath-taking, especially near Stevenson Island where a fissure or canyon was found to extend more than 400′ below the surface of the lake. Scientists were mesmerized by seeing underwater cliffs up to 75′ high, which they described as "a nearly vertical wall, stepped like some ancient fortress". The sheer drops were surprising not only because they occurred at depths much greater than any previously charted, but because they were fairly rocky or "lithified", in contrast to the soft and muddy sediments common in other areas of the lake.

Like all explorers, scientists and mariners, our researchers kept log books in which they noted the locations of their discoveries. This was done in 1987 and 1988 using a "MiniRanger Positioning System", which triangulated a position with two shore-based transmitters. For maximum accuracy, it was necessary for the younger members of the team to climb the same dead -- and of course, increasingly decaying -- trees each time the transmitters needed to be positioned. The group wanted to be absolutely sure of being able to revisit this newly discovered natural wonder under the lake of a park then being devastated by forest fires.

In 1989-90 team members were able to photograph the cliffs using a 35 mm still camera which had been added to the ROV's equipment, yet in 1992, even after completing some 30 dives over 12 days, they were unable to locate the spectacular cliffs. In what they thought was the correct location, all that could be seen were what looked like landslides or "sediment slumps": the cliffs had disappeared!

CONCERNING THE CHALLENGE OF NAVIGATION

Frustrated and puzzled, team members heard stories that in the fall of 1990, guests staying at the Lake Hotel reported seeing a "giant burp" out in the lake, which was followed by 5-10 minutes of visible roiling or disturbance in the water some distance offshore. It is known that significant earthquake tremors occurred near the lake after the 1990 summer season. Could either of these things provide clues to the disappearance of the deep canyons? Perhaps the clefts had collapsed inward, or simply been covered with silt as water flows changed. Our puzzled explorers took to referring to the lost cliffs as the "Walls of Jericho". They also wondered just a little if they were really looking in the exactly the right place...

Since 1989, the Global Positioning System (GPS) which interrogates satellites had been made available to the public by the Department of Defense. With a device no larger than a paperback book, a location could be identified within about 100′. The civilian GPS did not, in fact, yield as accurate a reading as the previously used MiniRanger system, but it was much more convenient aboard ship -- and no one had to climb trees! Even with this navigational aid, however, the explorers found no trace of the cliffs for three more years. In 1995 and 1996, the submarine's camera once again revealed rocky outcroppings reminiscent of those seen years before, but they were neither as steep nor as finely chiseled.

In 1989, a companion and I undertook the challenge of cross-country skiing several miles, carrying gear in to Cascade Lake in sub-zero temperatures. Our purpose was to install sediment traps on the lake bottom, equipment which would be collected for analysis the following summer, and hopefully yield valuable information about what occurred in the lake during Yellowstone's long, bitterly cold winter. Hauling and working with scientific gear under such daunting weather conditions was no one's idea of fun, yet it almost became so, because the struggle was not without purpose.

Scientific researchers seek to expand the sum total of human knowledge, a most demanding, yet glorious, endeavor.

Rob Paddock
UW-Milwaukee Great Lakes WATER Institute

Cliffs (above)
Rubble (below)

Steep cliffs...

apparently fallen to rubble.

Cliffs (above)
Rubble (below)

In 1986 we took a canoe trip up the Lewis river to Shoshone Lake in order to collect sediment cores as well as to investigate the possibility of doing ROV work on hydrothermal vents in Shoshone Lake. It was an arduous trip, which involved hand-towing canoes (which were loaded with supplies) up the Lewis River to the lake, a distance of some miles. Everything was natural with deadfalls of branches which we had to climb under or over, while wading in flowing water and sometimes falling into holes. I was certainly the oldest in the group, and at times was convinced that I would not make it and would simply be left by the side of the river to find my way back on my own.

Sectioning cores wearing mosquito netting.

PIPE DREAMS

Yellowstone is a legendary place of stunning contradictions, where steam swirls above ice and human visitors alternately flood the park and leave it nearly deserted. Our scientists' reports of their discoveries can sometimes reveal the ambiguities of human experience as well as dichotomies of nature.

One of the first underwater features unmasked using the ROV's (Remotely Operated Vehicle's) camera in Mary Bay was described as resembling a "pipe garden". Small, pale cylinders 3-4 centimeters (1.5 inches) in diameter and 15-25 centimeters (6-10 inches) high were observed protruding from the sediments. These turned out to be hollow tubes filled with soft mud, (presumably) relict "plumbing" or passages through which hot vent water had flowed into the lake. They were formed when hot, silica-rich vent water was suddenly cooled by contact with the cold lake, causing the silica to precipitate and form the tube walls. The resulting pipes were found to be quite hard -- breaking a piece off to sample was difficult.

CONCERNING HYDROTHERMAL PLUMBING

Nearly a decade later, in startling contrast to the small hard pipes, tall crumbly "spires" were discovered rising abruptly from an otherwise flat and monotonous area of the lake's floor near Bridge Bay. Lying in about 45 feet of water, a stone's throw from the northern shore of the lake, huge curved and crooked shapes towered to 20 feet high. While some measured 10 feet wide at the base and narrowed to a pointed top, other slimmer forms resembled large tree trunks.

These towers, like the small hard pipes of Mary Bay, are composed of silica, but they do not appear to have been formed within the sediments, as do the pipes. The spires, unlike the pipes, are not located near any known thermal area, nor do they appear to be emitting warm water. How and when they were formed is thus far a mystery. They somewhat resemble the Liberty Cap at Mammouth Hot Springs and the Soda Butte in the Lamar Valley, but these are composed of calcium carbonate precipitated in air, which is very different from the silica making up the tall towers and short pipes which lie underwater in Yellowstone Lake.

If you have found it confusing to read of objects described alternately in centimeters and inches, meters and feet, Celsius and Fahrenheit, you are not alone. Scientists who communicate and publish internationally use both English and metric instruments, and are so conversant with both that they switch back and forth without noticing the difference. For example, the Mini Ranger Positioning system described in the "Line of Sight" chapter calculates locations in meters, while the more sophisticated Global Positioning System now available identifies locations in both systems. Those of us not used to dealing with international measures on a daily basis appreciate reminders that those 15 centimeter pipes are about 6 inches long, while the 20 foot high towers measure just over 6 meters. Water freezes at 0 degrees Celsius and 32 degrees Fahrenheit, and it boils at 100 degrees Celsius or 212 degrees Fahrenheit -- at sea level, that is! In Yellowstone's higher altitude, with correspondingly less atmospheric pressure, water boils at 97 degrees Celsius or 200 degrees Fahrenheit.

Thus we see illustrated the challenge and fascination which is scientific exploration. Amazing things are painstakingly or accidentally discovered and minutely documented, only to unveil apparent contradictions and deeper mysteries. The physical universe continues to reward as well as humble those who attempt to search out its secrets.

Shoshone Lake is some seven miles across and we paddled from one side to the other, coring from both west and east ends. Later, while sectioning the cores, Bob Gresswell and I were attacked by thousands of mosquitoes – a horrendous experience! At times we felt that we were literally going to be picked up and taken away by the ferocious insects.

Our party of six spent the night camped in a ranger cabin on the shore of the lake. I am happy to report that our trip back the next day, down the river, was a blast. Tying canoes together, we rafted our way downstream on a joyride that was unforgettable.

Tony Remsen
UW-Milwaukee Great Lakes WATER Institute

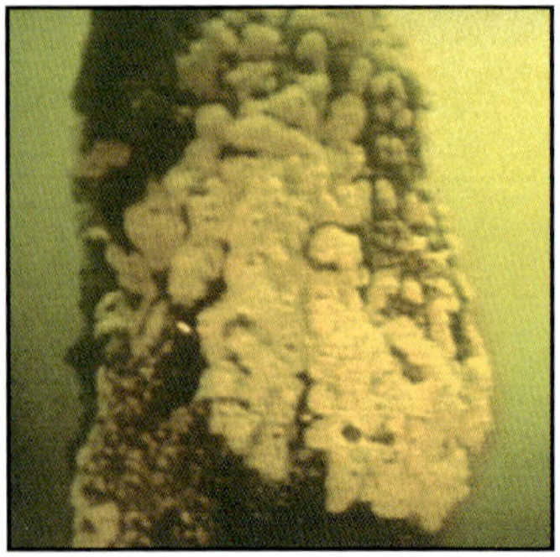

The 20′ Bridge Bay spires are host to sponges, snails and hydra.

Broken spire about 1 meter or 39 inches

Small pipe about 20 centimeters / 8 inches tall

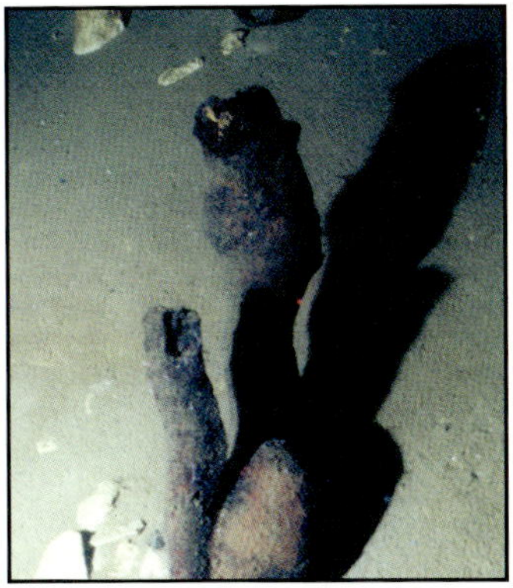

Relict hydrothermal plumbing lies exposed beneath Mary Bay in the "Pipe Garden" vent field.

Breakdown near Dunraven Pass

One of the moments I remember best about researching in Yellowstone actually occurred outside the Park. It must have been during the early 1990's; certainly during one of the early summers of our research, when we were working "on a shoestring" or very small budget. Because we didn't have the money to stay in hotels, we just drove all night across the Dakotas at 65 miles per hour. This particular night, as we pulled off the highway for gas, one of our trailer wheels seemed too hot, so Val Klump spit on the hubcap – which hissed! As we backed up, maneuvering into the station's repair shop, things got worse when the wheel first wobbled, then literally fell off!

You can imagine, we all breathed a sigh of relief when it occurred to us that this could easily have happened moments before at highway speeds.

bacterial mat

"SULFUR, ANYONE?"

Although located in the oldest national park, Yellowstone Lake remains full of surprises. Its waters harbor tiny organisms not typically found in deep alpine lakes, creatures so unusual and of such potential value as to be featured recently on national television news.

Who are these microscopic critters? They are bacteria of amazing diversity which feed off the nutrients flowing from the hydrothermal vents under the lake. Thanks to their prolific reproductive and communal proclivities, our scientists have been able to see these tiny beings with the naked eye since their earliest SCUBA reconnaissance dives in Sedge and Mary Bays in 1984. Some bacterial colonies appear as delicate coils of white filaments attached to plants, while others carpet the lake bottom as multicolored mats, which can become thick enough to trap gases beneath them. These organisms form the basis of the chemosynthetic food chain, metabolizing sulfur, methane and iron in order to grow.

Larger mats in shallower areas include photosynthetic bacteria as well. When many kinds of bacteria (which take advantage of differing environmental chemistries and temperatures) live in close proximity, the mats display a range of colors from white, to orange, green and purple. One such mat, seen in 1985, was 3 feet wide and close to 100 feet long as it undulated above a fissure in Sedge Bay.

CONCERNING LIFE AT HIGH TEMPERATURES

Fascinating as these creatures have seemed from the beginning of this research, their existence has recently taken on an urgent importance when coupled with explorations of the deep underwater canyons off Stevenson Island. Why? Because at underwater depths close to 400 feet, atmospheric pressure is greater (by about 176 pounds per square inch) than that at the surface, and prevents the water from boiling at the temperature it would on land. What has just been discovered is that, deep in these underwater clefts, water hotter than 100 degrees Centigrade (previously assumed to have been sterilized by intense heat) may in fact have bacteria living in it!

Because these organisms have the potential of tremendous commercial value to gene prospecting and bioremediation companies, the Park Service has taken steps to protect these living resources and to assure that if they are used commercially, a percentage of the profit will benefit Yellowstone Park and its visitors.

Besides the practical possibilities of benefiting our everyday lives, these newly discovered creatures may someday even give us clues about the origins of life on earth or other planets. On Europa (one of the moons of Saturn), hydrothermal activity may be occurring beneath a covering of ice, a situation which could be analogous to conditions beneath Yellowstone Lake.

As it turned out, we may have been safe from crashing, but were still in danger of being stranded, because the needed repair parts couldn't be flown into that remote location for a week. It was so hot, the temperature must have been in the 90's, close to 100 degrees. In desperation, we rented a gigantic Ryder truck (putting the bill on one of our personal credit cards, of course, since this change of plan was not included in our funding). I'm not sure what we would have done if three wonderful local kids hadn't offered to help us reload our gear from the broken trailer into the rental truck.

Splurging on ice-cold Cokes for all concerned, we then jumped into our various vehicles and drove like crazy in order to reach the east gate of Yellowstone Park before it closed for the night.

Jim Maki
Marquette University

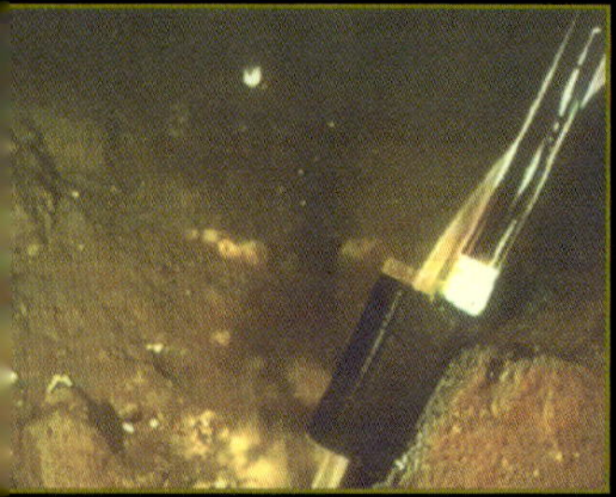

The ROV's probe samples a hot vent hundreds of feet beneath the surface of Yellowstone Lake.

Hot vents deep beneath the lake's surface are illuminated by the lights of the ROV.

I vividly remember one trip into Turbid Lake. On a very hot summer day, about six of us hauled coring equipment and two inflatable boats in along the old Turbid Lake Road. This was a struggle, since the road had been closed for some time due to bear activity in the area. We had to climb over and around fallen trees and other obstacles, carrying our gear and looking over our shoulders for the possible return of the bear which had treed two hikers only days before.

Once we managed to get back to the lake itself, our mission was to retrieve some sediment traps left there previously, as well as to take some fresh cores. These lake sediments are particularly interesting because they are distinctly layered, unlike those of most lakes which don't ice over in the winter (as Turbid doesn't, due to hydrothermal activity).

Overlooking Mary & Sedge Bays.

WE VALUE YOUR INPUT

As consumers, we are all familiar with trying to budget our money, to at least roughly predict our household income relative to expected expenses. Imagine the complexity of trying to "budget" the "income" and "expenses" of a lake! Such is the job of an aquatic biogeochemist. Some of the more obvious factors which affect lakes include:

~rivers and streams which flow into and out of the lake and what is carried with them
~the presence or absence of human development in the lake's watershed
~changes in climate and temperature (or events such as the 1988 fires in the Park)
~the life cycles of plants and animals which live in the lake
~exchanges with the atmosphere (evaporation and acid rain, for example).

Once scientists began to document the hydrothermal activity taking place under Yellowstone Lake, the fact that its water appears "richer" than that of most alpine lakes was (at least partially) explained.

CONCERNING NUTRIENT CYCLING

Yellowstone Lake turns out to be enjoying the benefit of a hidden "trust fund" of chemical wealth, which flows in from its underwater hot springs and fumaroles. Plumes of water enriched with carbon dioxide and radon can be observed, for example, streaming out into the lake for more than a mile from Mary Bay, one of the hottest spots in this unusual body of water. River water exiting the lake at Fishing Bridge has been found to contain 10 times more chloride, as well as higher concentrations of many other chemicals, than the water which flows into the lake from creeks, streams and the Yellowstone River in the lake's Southeast Arm.

The sediments under Yellowstone Lake are largely made up of discarded diatom (tiny plant) frustules, little relic "glass houses" which the plants construct from silica, a mineral found in the water of some hot springs in concentrations up to twenty times higher than that of open lake water.

Interestingly, not all springs emit the same chemical mix, though all their water is anoxic (or without oxygen) when it first emerges from underground and the overall concentrations of minerals rise consistently with higher water temperatures. Divers' surest visual sign of underwater geothermal activity are white granules of elemental sulfur which are deposited when chemosynthetic bacteria oxidize the sulfide escaping from the vents. (Gas samples collected underwater in funnels often reek with the rotten egg smell of hydrogen sulfide.) Vent water has been found to vary widely in its chloride content and can be acidic or of neutral pH.

Paddock & Klump still afloat in the RV Deathwish

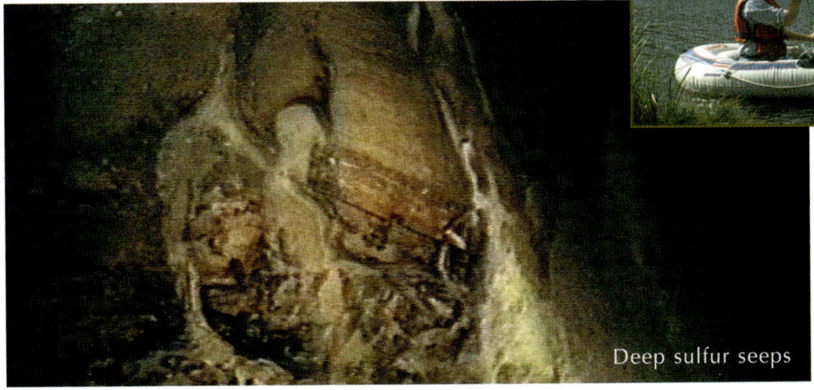

Deep sulfur seeps

Turbid Lake lies in an old explosion crater, and has a very deep hole (70-80 feet deep), as do Mary Bay and Indian Pond. We never managed to retrieve cores from the deepest spots, but did succeed in getting samples in about 12-15 feet of water. Working conditions were pretty unpleasant, not so much because of the heat as from the strong odor of sulfide which permeated the area, giving most of us headaches. We were coping pretty well anyway, until we discovered that the small inflatable boat we were using was leaking air. So there we were, pumping it back up with a foot pump in order to stay afloat and avoid sinking into water corrosive enough to dissolve stainless steel – nothing much lives in such water, other than acid-loving diatoms and bacteria. We had no desire to go for a swim, much less lose all our equipment!

The manufacturer ended up replacing our leaky boat with a new one which we subsequently referred to as the R/V Death Wish -- not that that kept us from using it, of course!

Pat Anderson
UW-Milwaukee Great Lakes WATER Institute

1983
Three scientists from UW-Milwaukee make their first trip to Yellowstone with a grant from the National Park Service.

1984-86
SCUBA exploration & collection of samples begins in Mary and Sedge Bays.

1985-86
Grant from National Geographic Society. Matt Kaplinski notes hot spots and steep gradients near Stevenson Island. A canoe trip is made up to the Lewis River to sample vents in Shoshone Lake.

1987-88
NURP funding brings Eastern Oceanic's Dave Lovalvo and the ROV to Yellowstone Lake for the first time. The ROV is deployed from the US Fish & Wildlife Service's boat, the Research Vessel Cutthroat.

Scene from the air before the fires

Forests begin to recover surprisingly quickly.

"Mini Ranger" positioning system is used to pinpoint locations in the lake. THe ROV is nearly stuck in the cliffs near Stevenson Island and the lake's depth is officially changed from 310′ to 390′.

1988

Fire in Yellowstone Park.

1989

Global Positioning System becomes available to civilians.

1990

Suggestion from naturalist John Dahlheim leads to filming the "trout jacuzzi" in West Thumb.

1992

Despite use of GPS and fathometer, steep cliffs near Stevenson Island can no longer be located.

1996

Spires are discovered near Bridge Bay at north end of Yellowstone Lake.

1998-99

Funding from the National Science Foundation.

2000

Funding from NURP and the Dreyfuss Foundation leads to publication of this book.

Worms head down in sediment.

We arose at 4:30 AM one morning to collect cores from a remote site. Following an all-day effort, our first successful sample was finally obtained at 4:30 PM. More collecting followed, despite a dramatic moment when a heavy, slippery, mud-covered core tube nearly pierced the bottom of our small research boat. Greatly relieved to have escaped having the boat sink out from under us, we returned to the field lab late in the evening to begin "sectioning" the cores, which always needs to be done as soon as possible.

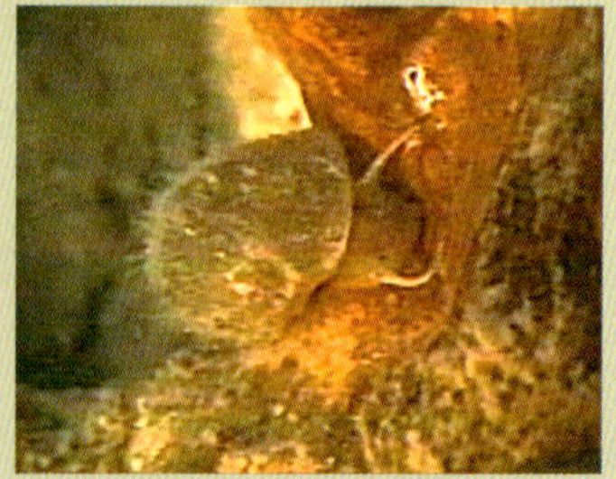

BOILED LEECHES

...are NOT on the menu at the Lake Hotel, though they do eventually get eaten, victims of their own gluttony. This is how it happens:

Attracted by the rich "soup" of bacteria and nutrients in the water emerging from the vents under Yellowstone Lake, animals and plants congregate and thrive near these hot springs in unusual numbers. "Autotrophs" (aquatic plants and some bacteria) are able to survive by synthesizing basic chemicals dissolved in the hot vent water into complex organic matter. The more complex "heterotrophic" organisms (e.g. worms and leeches) require compounds such as proteins and amino acids in their diets. Many of these life forms are too small to be seen with the naked eye, but others are perfectly visible to divers or on video screens linked to underwater cameras.

Most exotic, perhaps, are the hydra, stinging-celled relatives of sea anemonies. Snails, more familiar-looking predators, may be observed feeding on orange bacteria. Insects crawling along a submerged log look for all the world as if they were on land.

CONCERNING INVERTEBRATE ECOLOGY

The most numerous individually observed organisms are several small species of worms, including both nematodes (round worms) and oligochaetes (aquatic earthworms). They live head-down in the hot sediments near the vents, close to their thermal survival limits, with their tails protruding into the colder water above. Their nourishment comes from ingesting chemosynthetic bacteria. Higher concentrations of chemicals such as sulfide and ammonia in the vent water explains why the worms, like the bacteria they consume, cluster near the vents in numbers an order of magnitude greater than in the cooler sediments only inches away.

Also visible to the naked eye are the leeches which frequent these sites. Since, like the worms, leeches are cold-blooded, their metabolisms increase as they approach the warm vents, requiring more feeding to support their increased activity. The leeches become gluttonous predators of worms, their well-developed eyes and internal heat sensors helping them to hone in on their prey.

But, in the end, who eats whom? The leeches apparently cannot stand such hot temperatures, for many are found dead near the underwater fumaroles. Eventually decomposing, the dead leeches themselves become food for their intended victims, the worms.

Leeches too close to hot vents.

Joking about the dangers we'd encountered that day, Fred Binkowski and I were still working outdoors at 2 or 3 AM, when everyone else had gone to bed. Suddenly, a bear started growling in our immediate vicinity, near Hatchery Creek! Deciding that enough was enough, we finished the work inside, with doors and windows well closed, too tired from working for nearly 24 hours straight, to take any chances with a bear!

Jerry Kaster
UW-Milwaukee Great Lakes
WATER Institute

The second year we were out there, Jerry, Val and I were working down at Bay Beach marina, when it was getting to be about 5 o'clock. We were (overhauling the Research Vessel) Cutthroat's generator, when "Cow Bob" (US Fish and Wildlife) asked us what are we going to do that evening, because maybe we ought to try making some "Red Death". Not Val and Jerry's so much, but my eyes just lit up! Cow Bob did not know me, but I looked over at Val and Jerry and could see them kind of laughing about what Cow Bob was lucking into, as we agreed to go ahead. When we warned him that we had our own version of this, Bob's response was, "Yeah, but this is from the Southwest and they REALLY know how to eat spicy food down there!"

"We'll try it my way," I said. "And then we'll listen to some of your suggestions."

Well, we whipped up (my chili) and Cow Bob was convinced within seconds that the Southwest part of the US is not the only place where they eat spicy food!

The following year when we came out – we had maybe 4 people on that trip – we had a $300 budget for food. This was to cover everything, drinks and food for 10 days, it wasn't that much.

OLD TROUT

Back in the 1890's, when famous ichthyologist (fish zoologist) David Star Jordan determined that there were no fish living in Yellowstone Park's Lewis and Shoshone Lakes, a plan was devised to stock the two lakes with Lake Trout or "Mackinaw", a kind of fish which came originally from Scotland or Germany. Accordingly, spawning trout were collected in the fall of the year from Lake Michigan, because it, like Lewis and Shoshone, was cold, clean and clear. Harvested eggs were then artificially fertilized, incubated over the winter, then hatched and raised to fingerling size in a hatchery. The following August, about 40,000 fingerlings -- along with 40,000 fingerling Brown Trout to serve as forage fish -- were transported in milk cans, by train, mule and pack horse all the way across country to their new home in Yellowstone. There the fish thrived, along with a later addition of red-sided or Utah Chub, which were also introduced as forage fish for the Lake Trout.

Around the turn of the century, park naturalists noted that the trout had begun to reproduce naturally in Lewis and Shoshone Lakes, on their way to becoming, in ichthyologist Fred Binkowski's words, the "King of Beasts" in Yellowstone lakes. "Every place they put the Lake Trout, it did well."

Except, perhaps, in Lake Michigan itself, where sometime shortly after World War II, in the late 1940's, Lake Michigan's own population of Lake Trout collapsed and stopped reproducing in the wild. Since that time, the trout, ironically, have had to be raised in hatcheries and stocked in the Great Lakes.

Not surprisingly, Fred Binkowski and Jerry Kaster, as a members of the Wisconsin research team, wondered why (presumably) genetically identical fish would continue to reproduce in Lewis Lake, while becoming unable to do so in Lake Michigan. So far, two main approaches have been used to investigate this mystery.

The first involved coring from Lewis and Shoshone Lakes, since a lake's biological history can be discerned from a study of its sediments. Knowing that no fish existed in the lakes prior to the 1890's, as well as the approximate dates of stocking and the onset of natural reproduction, scientists could analyze the environmental significance of these events through changes evidenced in the sediment cores.

Binkowski and Kaster's second, more high-tech approach consisted of video-taping spawning fish in the wild:

"We (got to know) the Lake Trout spawning areas. The NE corner of Lewis Lake is a shallow area with lots of rocky rubble. So we went back in the fall, I think about 3 years in a row, and used an underwater black and white camera. We were able to do so quite effectively just by lowering it on ropes like a puppet, because we were working in maybe 5 to 7 feet of water most of the time. (The pictures fed) back into the monitor that we had on the boat. We always did this at night and there were just tons of fish. We could see the fish down there because of the lights on the camera, you know. We found fish galore on these reefs. They were spawning, the females were dropping their eggs, the males were coming up alongside and doing the fertilizing as the eggs were deposited and dropping down into the honeycomb rock. It was just incredible!"

Since the scientists were already familiar with the seasonal, temperature and photo period requirements for spawning, the video observation allowed them to learn exactly what types of substrates were ideally suited to the purpose, and to observe behavior patterns. "They would congregate in big groups and at certain times. For example, there weren't as many fish at 9 o'clock as there would be at midnight."

Much as Binkowski and Kaster were able to learn, however, the problem of Lake Trout reproduction in Lake Michigan remains a mystery. The two friends have many ideas for further experimentation which include:

~ Replicating the 1890's stocking to see if fingerling domestic fish from Wisconsin would reproduce naturally in Wyoming. Binkowski points out that this experiment would require much patience, as the fish do not begin to reproduce until they are 7 years old.

~ Bringing wild fingerlings from Yellowstone to Wisconsin to see if the wild fish would continue to reproduce naturally in the larger, more easterly Lake Michigan from which they came. Of course, this too would require the long 7 year time span.

~ Taking advantage of recent improvements in DNA fingerprinting, the scientists would like to compare genetic changes or similarities in the Wisconsin and Wyoming populations, as well as compare them with trout which have recently begun to reproduce naturally in Lake Superior.

We stopped in Cody, Wyoming and the guys sent me into the store – that was their first mistake – to do the grocery shopping. (As it turned out), a third of the budget was dedicated for chili ingredients to make what is now referred to as the RED DEATH.

We used Cow Bob's cabin when he was gone for the day. (When) he came back that evening, everything was cooking, the smells were penetrating the air. He'd probably smelled it from outside already, (because) he walked in the front door (and just) made a bee line for his bedroom, saying "I'm not eating any of that stuff!" He'd learned from the first trip!

(Editor's note: Fred made so much chili that his companions had to eat it at every meal for three days!)

Fred Binkowski
UW-Milwaukee Great Lakes
WATER Institute

Special Thanks to
Our Many Collaborators,
Students and
Yellowstone
Friends:

Glen Boltz
Dan Carty
Brian Ertel
Bob Evanoff
Rick Fey
Bob Gresswell
Ron Jones
Harlan Kredit
Dan Mahony
Dan Reinhart
Jim Ruzychi

We gratefully acknowledge the financial and "in-kind" support of the following agencies, without which the research and publication of this book would not have been possible.

National Undersea Research Program

National Park Service

U.S. Fish & Wildlife Service

National Science Foundation

UW-Milwaukee Great Lakes WATER Institute

UW-Milwakee Graduate School

Marquette University

National Geographic Society

The Dreyfuss Foundation

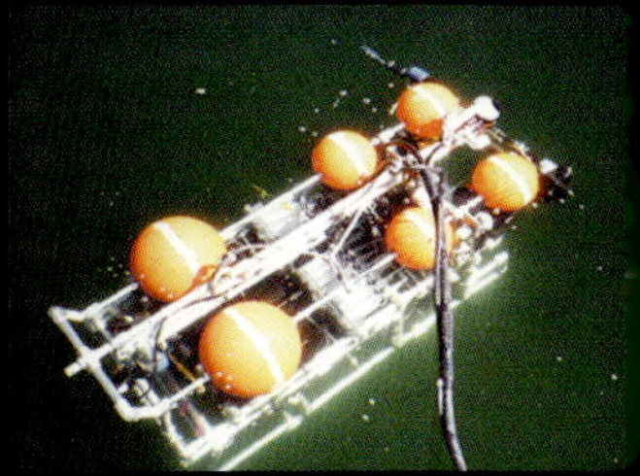

YELLOWSTONE LAKE:
JOIN THE EXPEDITION!